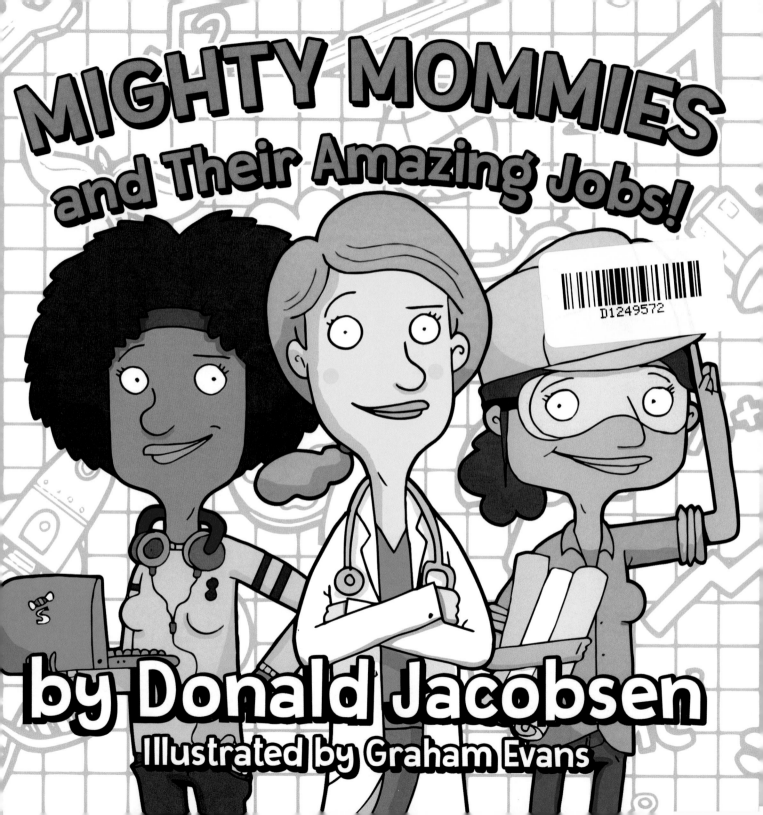

MIGHTY MOMMIES
and Their Amazing Jobs!

by Donald Jacobsen
Illustrated by Graham Evans

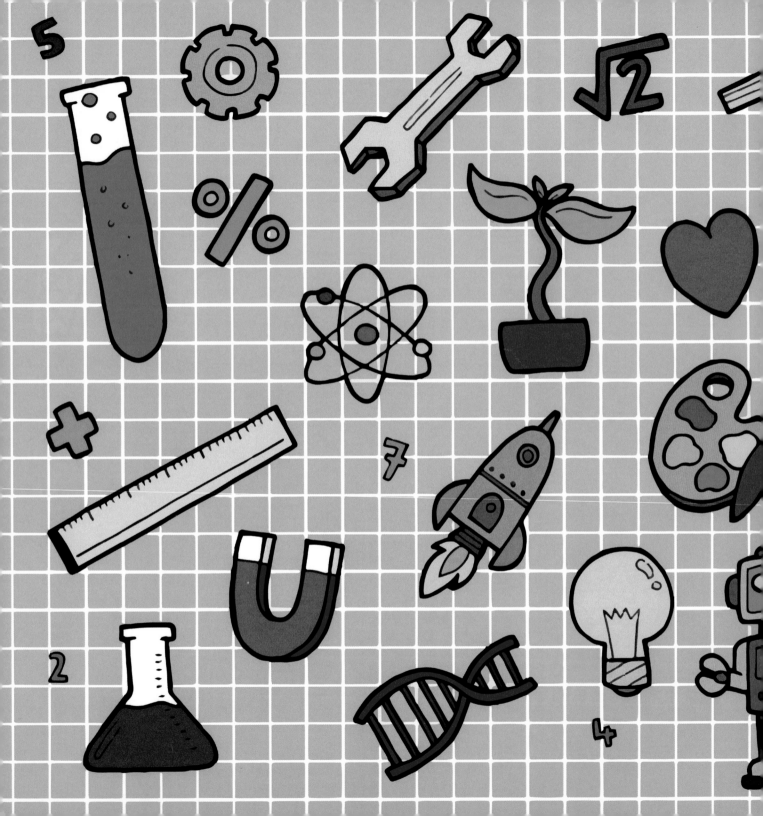

Mighty Mommies and Their Amazing Jobs
A STEM Career Book for Kids
STEMpowering STEM Books for Children Series #1
by Donald Jacobsen
Illustrated by Graham Evans

Second Edition Printed in the United States of America

an imprint of Three Suns Press
Memphis, Tennessee

ISBN (paperback) 978-1-7328273-2-5
ISBN (hardcover) 978-1-7328273-1-8
Library of Congress Control Number: 2019901367

BISAC Subjects:
1. JUV006000 JUVENILE FICTION / Business, Careers, Occupations
2. JUV014000 JUVENILE FICTION / Girls & Women

This book belongs to:

JANSI

LATHIA

Daniel's mommy is a

Doctor.

She treats people who are sick.

She looks at patients' X-rays and helps them heal real quick!

Ava's mommy is an Architect.

She draws blueprints for construction.

Builders carefully read her plans and follow her instructions.

Parker's mommy is a
Pilot.

Flying airplanes is her skill.

She travels all around the world from Australia to Brazil.

Mia's mommy is a

Marine Biologist.

She studies life in the ocean.

She examines coral and turtles and fish,

and her work takes a lot of devotion.

Peyton's mommy is a
Police Officer.

She helps keep us protected.

She's friends with everyone in town

and she's very well respected.

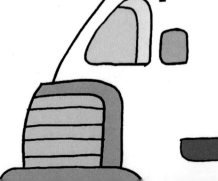

Paisley's mommy is a

Programmer.

She works on a computer.

If you need help writing code,

she'd make an awesome tutor!

Eli's mommy is an Engineer.

Fixing things is her craft.

She can build all kinds of high tech stuff,

starting with a simple draft.

Tara's mommy is a Teacher.

She carefully plans her lessons.

She works with kids and parents and helps answer all their questions.

Aiden's mommy is an **Attorney.**

She knows all about the law.

She works with judges and police officers to make sure everything's fair for all.

Paige's mommy is a Pharmacist.

She helps her patients get well.

She knows all about different medicines used to fight off bad germ cells.

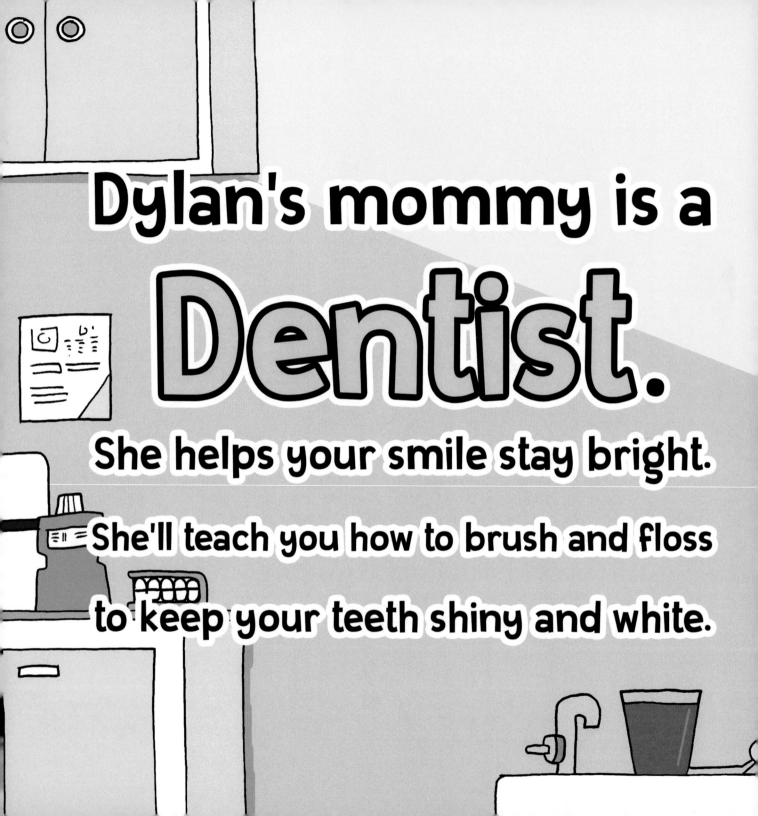

Dylan's mommy is a

Dentist.

She helps your smile stay bright.

She'll teach you how to brush and floss

to keep your teeth shiny and white.

Faith's mommy is a **Firefighter.**
She's there to save the day!
Call for help if you see a fire
and she'll come right away!

Sam's mommy is a

Scientist.

Running experiments is her quest.

When it comes to solving difficult problems,

she's the very best.

Charlotte's mommy is a

Chef.

Her job uses science and art.

She can cook anything from scratch,

from soups to pastries and tarts.

Paul's mommy is a Paleontologist.

She studies fossils and bones.

One day she hopes to open a zoo full of dinosaur clones!

Aubrey's mommy is an
Accountant.

She works with numbers and charts.

She runs a busy business

and she's very detailed and smart!

Adam's mommy is an

Astronaut.

Working in space is her thing.

She's setting up a giant telescope

that can observe Saturn's rings.

Vera's mommy is a

Veterinarian.

She helps animals at the zoo.

From koalas to monkeys to lions and tigers,

there's plenty of work to do!

There are so many MIGHTY MOMMIES with lots of jobs to do!

From writing a program to working a case, there's just so much to choose!

FREE audiobook and resources for parents and teachers!

Because you play such a challenging and meaningful role in the development of children, this book includes a free audiobook and lesson plan resources.

Go to the link below to get access to your free bonuses!

www.donaldjacobsen.com/mmaudio

If you loved this book, please leave a review on Amazon. Reviews are the best thank-you that you can give an author!

About the Author

Donald Jacobsen is a dad, husband, registered nurse, and sometimes-writer living and working in Memphis, Tennessee. When not busy working on his next book, he enjoys spending time with his wife, two daughters, and their mopey rescue dog, Yoda.

Learn more at www.donaldjacobsen.com.

To my three sunshines, Stephanie, Hazel, and Holly.

Special thanks to Debra Hoffman and Megan Silea, who assisted with the editing of the first edition and who also sparked the title of this book. Many thanks to all the friends and family who continue to believe in me and inspire me every single day.

-Donald

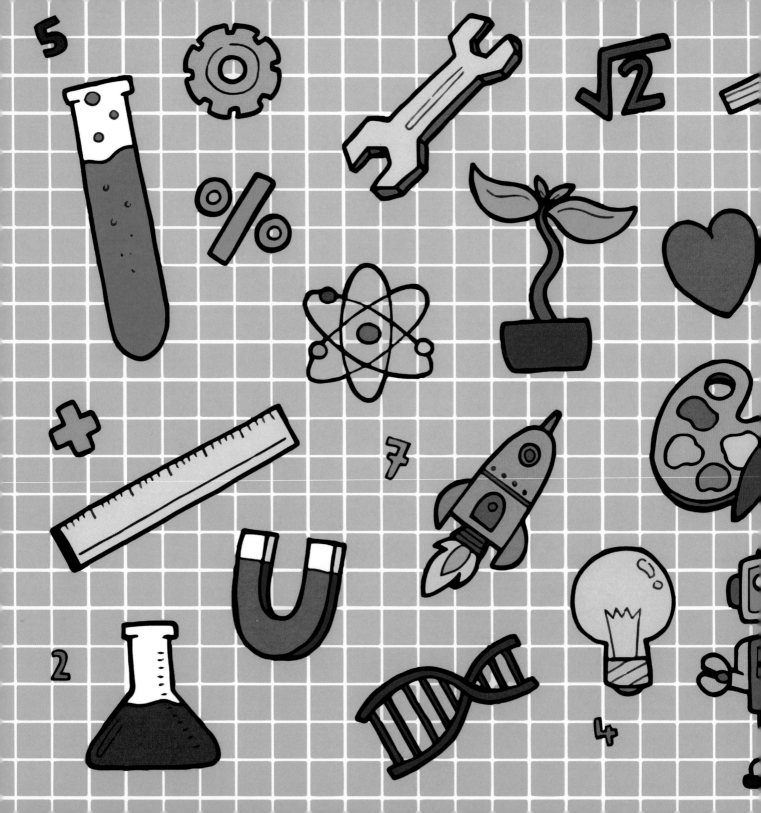

Made in the USA
Middletown, DE
19 July 2019